D0775199

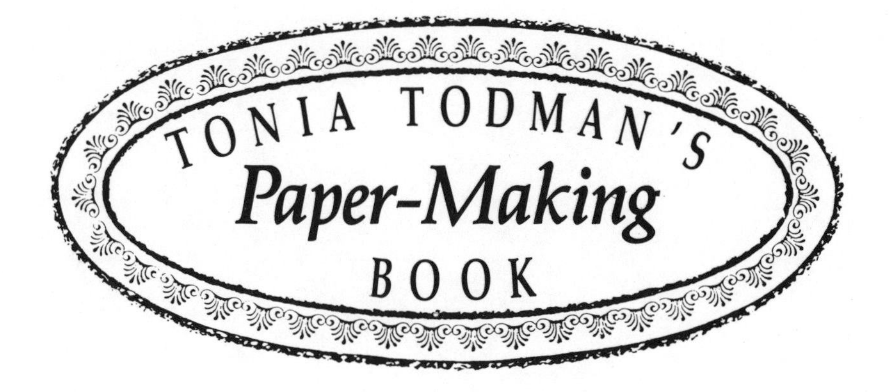
TONIA TODMAN'S
Paper-Making
BOOK

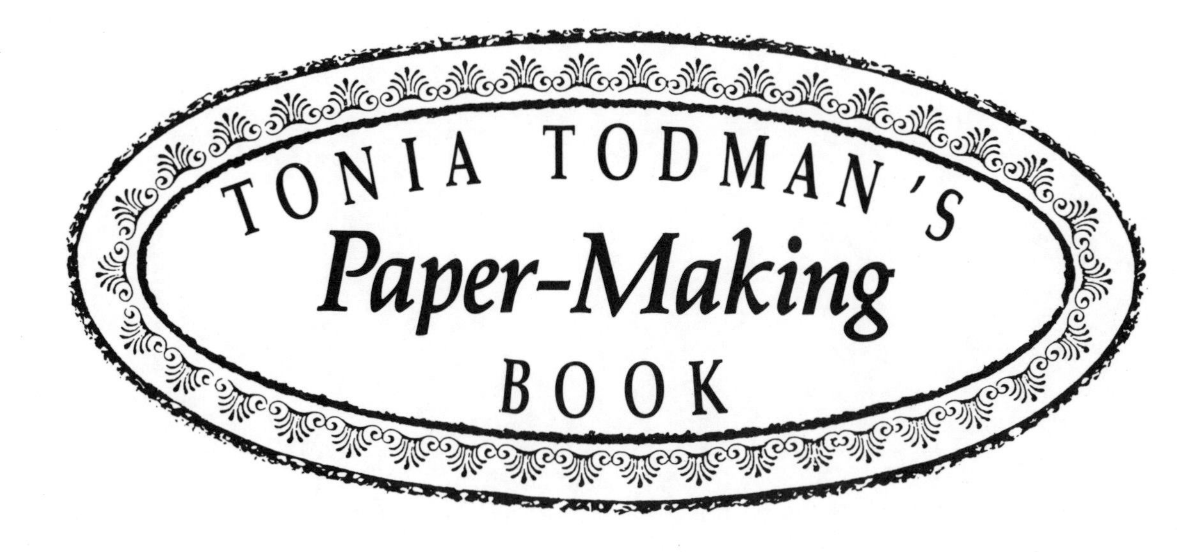

SALLY MILNER PUBLISHING

First published in 1992 by
Sally Milner Publishing
558 Darling Street
Rozelle NSW 2039
Australia

Reprinted 1992, 1993 (twice)

Design by Gatya Kelly
Layout by Shirley Peters
Photography by Andrew Elton
Illustrations by Angela Downes
Typeset by Shirley Peters
Printed in Malaysia by SRM Production Services

National Library of Australia
Cataloguing-in-Publication data:

Todman, Tonia.
Tonia Todman's paper-making book.

ISBN 1 86351 097 4

1. Papermaking. I. Title. II. Title: Paper-making book.

676.22

Contents

Introduction 1

The history of paper making 2

Some good reasons to make paper 4

Material and equipment 6

To make pulp 6
How to make a mould and deckle 8
A mould and deckle for envelopes 10
The vat 11
How to make a couching pad 12
How to make a paper press 15

Making paper pulp 17

What papers are good for recycling 17
Making pulp from plant fibre 21

Making sheets of paper 23

Your workplace 23
Pulling sheets of paper 24
Couching the paper 26
Pressing the paper 28
Separating paper from couching cloth 29

Common problems 30

Paper is too thin 30
Paper is too thick 30
Paper splits through the centre, or at its edges 30
Weak paper 31
Pulp lumps 31

Paper will not separate from couching cloths 31
Stains in paper 32
Spots of solid colour, and transparent spots 32
Dry paper has creases, or indentations 33
Straightening wrinkled dry paper 33

Creative effects **34**

Additions to the pulp 34
Paper patchwork 37
Simple marbling 38
Creative suggestions for using marbled paper 39
Stencilling your handmade paper 40

Glossary for paper making **42**

Projects shown on colour pages **46**

Introduction

Paper making is one of those lovely, undemanding crafts that engrosses you completely. It is perfect for the unskilled, though those of you who wish to become master paper makers can easily acquire those extra, specialised skills. Paper making costs you very little, there is no lingering accumulation of your output and, apart from a few easily learned techniques, does not require perfection. It's a terrific craft activity for children, too. They can have a lot of fun with paper making, and it's a sure confidence booster to those of you who would swear to being totally uncreative!

I've written from personal experience, and couple this with traditional methods and advice from master paper makers. The techniques mainly use recycled paper, so the important issue of preserving our natural forests is being addressed – albeit in a very small way!

I hope you have as much fun as I do while I'm making paper.

The history of paper making

Paper, as we know it, is a very modern invention when compared to the amount of time civilised man has lived on Mother Earth. Before the art of making paper making was discovered there were few surfaces suitable for writing, clay tablets, silk cloth and stone fragments being the most available. People had learnt how to fashion vellum from calf skin, and parchment from sheep skin, and the Egyptians had papyrus which was the first real paper as we know it. Despite attempts by the Egyptians to prevent the outside world knowing too much about papyrus paper making, the word had spread and it was soon in use throughout Asia Minor and wherever the Greek and Roman Empires had influence.

Papyrus is a reed-like plant and grows in or very near water. The Egyptians split the stems, removed the green outer layer, then laid the inner part of the reeds side by side onto stone. Another layer was placed over the first, in the opposite direction. More layers were built up and kept damp, then hammered until the whole lot had compressed down into a tablet form. River stones were rubbed over the surface to smooth it and to give the papyrus lustre. The oldest fragments yet found are archeologically dated as being 4500 years old!

The accurate origins of the discovery of vellum and parchment are lost in legendary stories, but it is fair to say that the Greek and Roman empires had a firm grasp on their production. These two materials remained important for official documents both from the church and government, for the production of complete books. Even when paper making was more common in the 12th century, vellum and parchment were still regarded as more long lasting, and therefore considered the only real choice for religious and legal documents.

It is said that a Chinese court official, T'sai Lun, first discovered the art of making paper from plant fibres, and that he 'patented' his invention around 105 AD.

Archeological evidence suggests that T'sai Lun may have in fact borrowed one or two ideas from existing forms of primitive paper made hundreds of years before he made his first attempt! Nevertheless, the Chinese worshipped him as a God in recognition of his achievement, and paper making as we know it became yet another invention attributed to the creative Chinese.

T'sai Lun had taken old fishing nets and ropes and beaten them to a pulp in running water. He then spread the pulp over a screen of closely tied bamboo strips, which acted as a sieve, allowing the water from the pulp to fall away, while still holding the fibres of the paper pulp together. He pressed the pulp by placing a weight upon it, and left it to dry. When the pulp had dried, the 'paper' was able to be peeled away from the bamboo base.

The Chinese were innovative in the materials they employed to make paper pulp, and they were the first to use rags for paper. Some fragments of rag based paper, dated to around the same time as T'sai Lun's first attempts, have been uncovered by archaeologists in the Great Wall. Silk fibres were used extensively, but soon became a luxury when the use of plant fibres in paper making became common.

The Chinese were very much a self-contained people, sharing little of their ideas and culture with the outside world. It was not until 755 AD at the siege of Samarkand when, we are told, Arabs captured some Chinese paper makers that their paper making secrets were learnt. Naturally, these methods were welcomed at home and it was not long before paper making spread throughout the Mediterranean.

Japan and Korea had started making paper soon after the Chinese, and Japan is still regarded as the source of some of the finest quality handmade papers available. Japanese are credited with having made paper so strong yet flexible that it could be used to make both clothing and household walls! In its travels across the Asian continent, papermaking became well established in India, where it is still an active village enterprise.

The first paper mill was established in Spain in 1150, and in 1490 the first English paper mill was established. In 1688 paper making was started commercially in America.

In times gone by the method of making paper from rags was singularly unpleasant, and decidedly unhealthy.

Rags were allowed to break down in water for weeks and then were chopped into tiny pieces – usually by women and children. Needless to say these vats full of rotting rags would have been amongst the last places you would wish to work! After chopping, the rags were taken to a stamping mill where they were pulped by large, heavy mechanical stampers, each pulverising rags contained in smaller vats, helped along by the use of copious quantities of running water.

As time went on, the whole process became somewhat mechanised. Rollers and beaters were used to run over rags left on conveyer belts, running water rinsed the pulp and more rollers took over to flatten out and eventually form the pulp into paper.

While books were largely out of the reach of common people – as most of them couldn't read – and books were being written by hand, the demand for rags was equal to the production of books. The invention of the printing press had an enormous, and inevitable effect on the demand for paper. Suddenly England had to import scrap rags from Europe. I'm told that some historians believe the Black Death plague in England resulted from infected rags destined for paper making being imported from Europe.

It was the nineteenth century before wood was used as the main fibre in paper making. Rags are still used today, but only by specialist manufacturers and master paper makers. Interestingly, artists still prefer to paint their masterpieces on wood-free paper, as the acid chemicals used in woodpulp make the paper less enduring than paper made with rags.

Some good reasons to make paper

The aspect of recycling has to be one of the main reasons for making paper. We all know that the world's natural resources are dwindling rapidly, and that any gesture towards saving or recycling resources is a good one. The mountains of scrap paper available daily from some offices, and other piles of paper accumulated regularly in average households must make us all think, and I'm sure we all want to reduce our consumption of the purchased product, though I'm the first to suggest that this is not easy! At best we can all make an attempt to conserve and re-use, but if we extend those thoughts just a little you can see the creative opportunities.

Imagine all the household scrap paper, and maybe some from the office, converted into lovely writing papers and envelopes, invitations, paper for drawing upon, or small pieces for greeting or business cards. Suddenly conservation becomes a very attractive pastime.

Paper making doesn't cost a lot of money. Some other crafts require expensive ingredients, or machinery, and this can inhibit people from learning or participating. The raw materials usually cost nothing, and if the end result – your paper – is not to your liking you simply pulp it again and re-use it. The moulds, deckles and couches can all be simple, or improvised, and the workplace for paper making can be mostly outside the home.

Many crafts naturally lead to an accumulation of things produced. Paper making produces materials that are in demand, and I think it's safe to say that you'll be pushed to produce more paper than you can use. It is off being used before you know it, and when its use is over, it's able to be recycled, again!

Paper making has no age barriers. It is a gentle craft, requiring no great physical strength or ability. There will be some skills learnt along the way, but, generally, it is absolutely right for the beginner, child and adult alike. Those who would normally call themselves uncreative can get great satisfaction from paper making – even the simplest results are exhilarating. And for those who would like to forget their troubles, it has been said that you have to concentrate so much when making paper that other worries disappear!

It's a good children's craft, too. Children take to this craft quickly, and you only really need to scale down the height of the workplace to make it accessible. Results don't have to be perfect, so it will easily become a confidence-boosting pastime.

Try not to think of your handmade paper as a luxury – though I do when I see some especially beautiful pieces. Batches of paper and envelopes make beautiful presents, and you should encourage the receiver to use them, not hoard them away. If you are actively involved in fund raising, paper is a good seller and an enjoyable activity for your group of workers.

Material and equipment

To make pulp

From waste paper

- Scrap paper torn to size (for more information see *Making paper pulp from paper*)
- Buckets, garbage bins or basins to hold the pulp
- A kitchen blender, or if working with large quantities of paper, an electric drill with a long attachment for mixing paint
- A kitchen sieve

From plant material

- Suitable plant material, mainly leaves and soft stems (see more information under *Making pulp from plant fibre*)

- A pot for cooking the plant material, and a stove top
- Caustic soda
- A kitchen sieve

To make paper from the pulp

- A mould and deckle (instructions on how to make these follow)
- A vat, which can be a basin, similar to a baby's bath, which is oblong shaped and larger than your mould and deckle
- Quantity of paper pulp, or plant pulp, and water
- A couching pad to rest the newly formed paper on. This can be made up from a large lipped tray, a piece of foam sponge to fit within the tray and a piece of discarded blanket or similar woollen fabric to fit within the tray, and newspapers; or a simpler version involving blankets and your paper press. Instructions follow in the book about making the various couching pads
- Couching cloths, either synthetic fabric kitchen wipes or pieces of cotton fabric (calico or old sheeting) about 40 x 50 cm (16″ x 20″). I prefer to use cotton fabric as the indentations left in the paper are less obtrusive than those of the kitchen wipes
- A strip of cotton fabric the same length as your mould and 10 cm (4″) wide for your couching guide marking cloth
- Pressing boards, either heavy bread boards, or a paper press (hard to find, so see more information later about making your own paper press that will require 40 x 50 cm (16″ x 20″) pieces of plywood)

How to make a mould and deckle

By definition, your mould is just a simple rectangular wooden frame with mesh stretched then fixed across it.

The deckle is an open frame of exactly the same size as the mould, with no mesh. The deckle will lay across the mould on the mesh side, matching edges exactly and turn the two pieces into a sieve. The deckle should be a little thinner than the mould if possible, as the depth of the deckle helps dictate the thickness of your paper. Try to make your deckle about 1 to 1.5 cm ($^1/_2$" to $^5/_8$") thick, but the outside and inside dimensions should be identical to the mould.

The size of your mould and deckle dictate the size of your paper or envelope. While most sizes are possible, I suggest you start with a mould no larger than A4 paper size. My first mould some years ago was about the size of a standard wedding invitation, considerably smaller than A4 size. The inside measurement of your mould and deckle for A4 sized paper is approximately 30 x 21 cm ($11^3/_4$" x $8^1/_4$").

The wood used for your mould and deckle needs to be planed and of the non-staining type, such as pine or light coloured hardwood. You will often find carpenters'

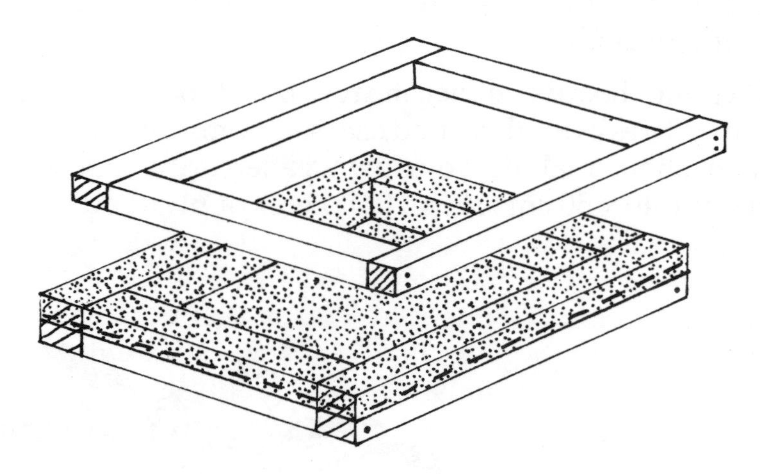

The upper frame is a deckle, the lower is a mould

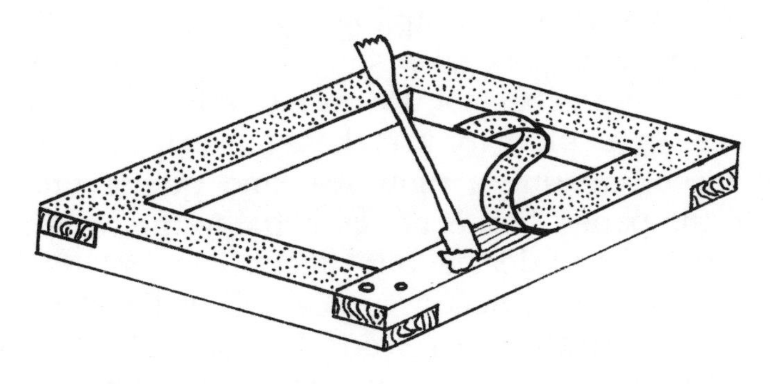

Gluing felt strip to the deckle base

off-cuts available that are most suitable. I suggest you look for wood that is 3.5 x 2 cm (1³/₈″ x 1″). The corners could be dovetailed together if you are an accomplished carpenter, though simple butted corners are perfectly adequate. Please use brass tacks, staples or galvanised nails to fasten as they won't rust. Corner plates screwed with brass screws to the outside edges will help reinforce the joints. Water will have a weakening effect on these joints, so the addition of a little wood adhesive to the edges before joining will help. Further information about a mould and deckle for envelopes appears later in the book, but the method for making these is identical to that for moulds and deckles for paper.

After you have made your two matching frames, one of them has to be covered with mesh to make the mould. Nylon mesh, sometimes sold as flywire, is suitable, though experts like to use fine bronze wire mesh. The smaller the holes are in the mesh, the better. Experts like to use mesh that has 12 to 20 holes per centimetre (30 to 50 holes per inch). Some fabrics are also suitable, and I have had great success with a heavy-duty cheese cloth. Some lace curtain fabrics are suitable, though remember that nylon stretches when wet. Stretch the mesh or fabric over the top of one frame, holding it as taut as possible. This is best achieved by tacking or stapling along one long side (not on the top of the frame) first, then stretching the mesh across the frame to take it to the opposite side. Then tack the short ends, using the same sequence of steps.

The most important thing to achieve with mesh is tautness, and while some slackness will not affect your paper, you will find it easier to remove the pulp from a

taut mould, rather than a slack one. If you are using fabric as mesh, wet it before stapling or tacking it in place.

A traditional paper maker sometimes places strips of wool felt along the base of the deckle. When the mould and deckle are in place for use, these felt strips sit on top of the mesh, creating a tight seal and giving precise edges to the paper. This is not essential to the workings of your mould and deckle, but I feel it is an interesting addition to the paper making process. Glue the strips in place using a strong PVA wood-working glue, being sure it is totally dry before immersing it in water.

A mould and deckle for envelopes

Use a stiff piece of cardboard to help you shape envelopes

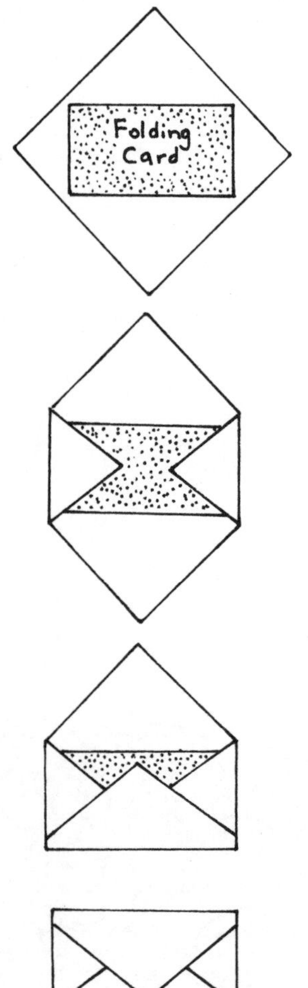

Envelopes should be made to accommodate the paper you pull. The mould and deckle is made in exactly the same way as a paper mould and deckle, and the paper is processed in the same way. The only difference is its shape. It is possible to make a complicated mould and deckle for your envelopes, involving curved lines and small indentations. This will probably frighten off beginners! I would opt instead for a simple 'squashed' diamond shape, that allows you to shape a classic envelope. Or, to be even more simple, make a square mould and deckle and then fold the paper, as shown, into an envelope shape. Or, simpler still, place a ruler along a sheet of A4 paper and tear a 9 cm (3½") strip from one end. This leaves you with a 21 cm (8¼") square from which to fold your envelope. Practise folding your envelope on scrap paper first, so that you do not waste paper with incorrect folds. Glue your envelopes together using PVA glue, applied with a small paintbrush.

If you receive an envelope shaped in an attractive way that appeals to you, undo it and explore the possibilities of making a mould and deckle of that shape. Remember, though, that it may be difficult!

Stencilled writing paper

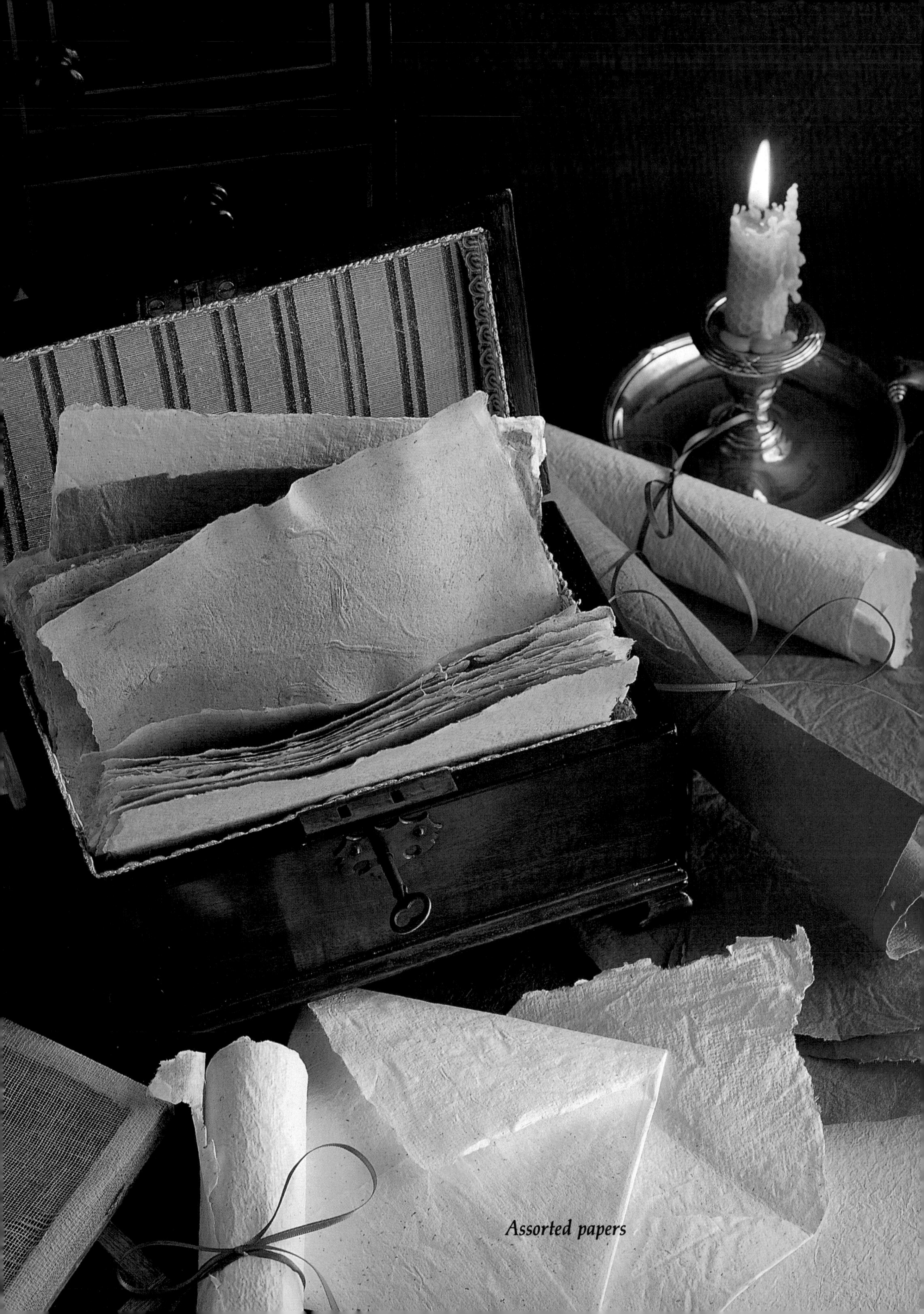

Assorted papers

Ingredients for paper making

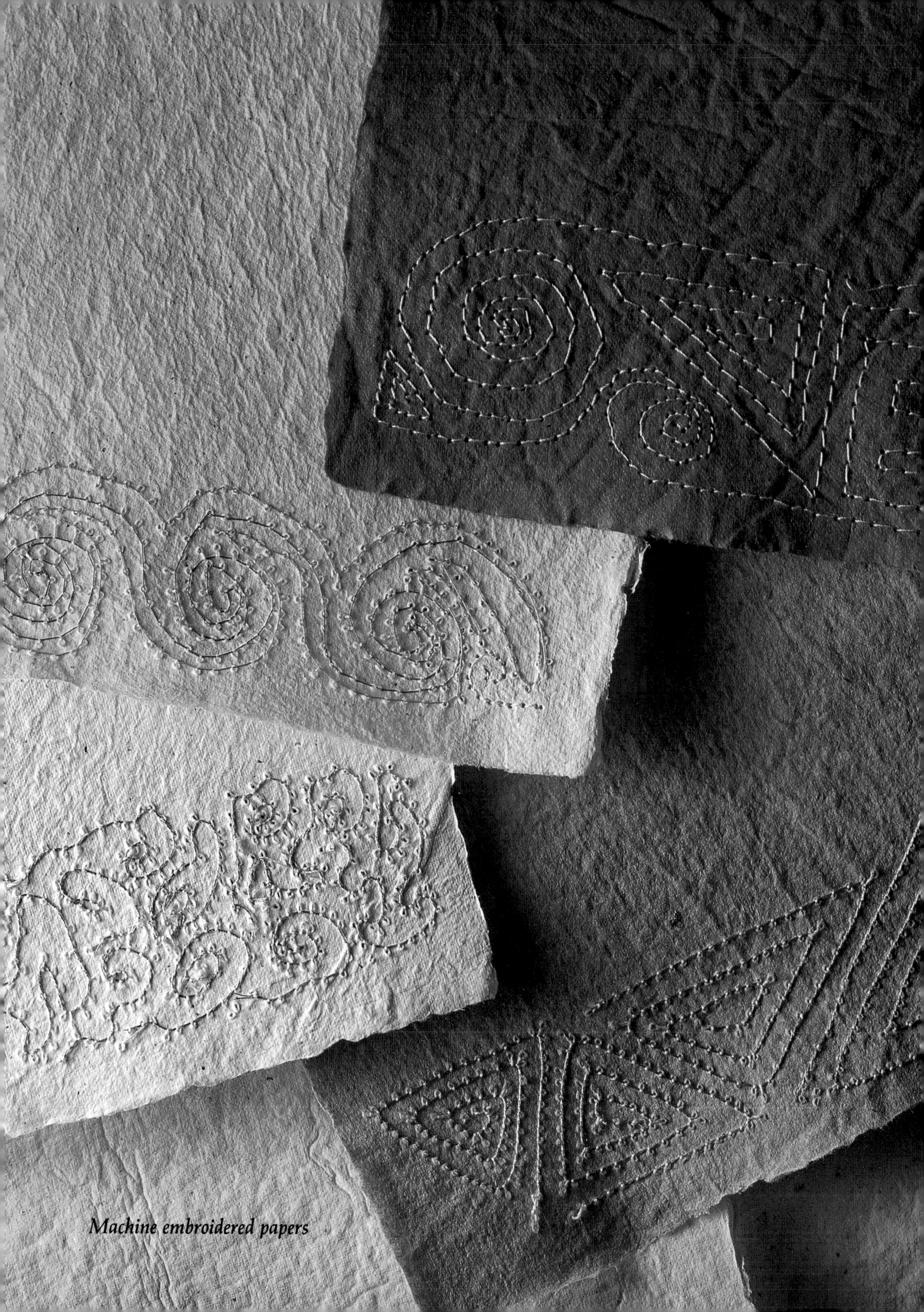

Machine embroidered papers

Monkey wrench paper patchwork

Plant paper patchwork

Using a mould and deckle

A watermark

Couch and couching cloths

A pot of flax

The vat

The vat is a functional element in paper making. It is the vat that holds the liquid pulp, into which you scoop your mould and deckle to collect paper fibre. The vat should not be so deep that it is awkward for you to work at, and be mindful of the height of your work bench, considering the height of the vat sides. The vat must be bigger than your mould and deckle and spacious enough for you to be comfortably able to sweep the mould and deckle through the pulp and out the other side. Any hindrance to this smooth motion will result in distorted paper. If you wish, you could work in spacious laundry tubs, sitting your vat inside one of the bowls, then any spillage could be easily washed away. If using very liquid pulp, you could actually use the kitchen sink as a vat, with the drainage boards being the ideal spot for your couching pad. The mixture of paper fibre and water is so liquid that any spillage will not clog up your drains. In fact, it would be finer than the liquid resulting from sink-installed food scrap pulverisers. There is so little fibre left in the sink at the end of your paper making session that you are virtually draining away only water.

Even though any spillage is going to be water, it is a good idea to have a drip tray under your vat, or at least a wad of newspapers that can absorb the paper pulp. Your enthusiasm soon takes over with paper making, and you will resent having to stop and mop up!

Vats can be baby baths, the kitchen sink, commercial storage bins or even a very substantial box or carton lined with thick plastic.

There is no reason not to use warm water in your vat during winter – it's so much more comfortable! You may even prefer to work wearing rubber gloves.

How to make a couching pad

The couching pad is where your fresh paper pulp is laid to dry. It needs to be portable, and you will need to position it where water spillage will not cause problems. I will describe two ways of making up a couching pad. Both are efficient, though method 2 has less handling of the wet paper, and could be simpler for children to construct and use. Both involve similar materials, and both need marker cloths.

METHOD 1

You need as the inner core of your pad something absorbent, which will stay wet, and I recommend plastic foam, though you could easily substitute cotton towels, or an old blanket. The foam, or folded blankets or towels, need to be about 10 cm (4") bigger than your paper all round, and as mentioned in the equipment list, it needs to fit within your lipped tray.

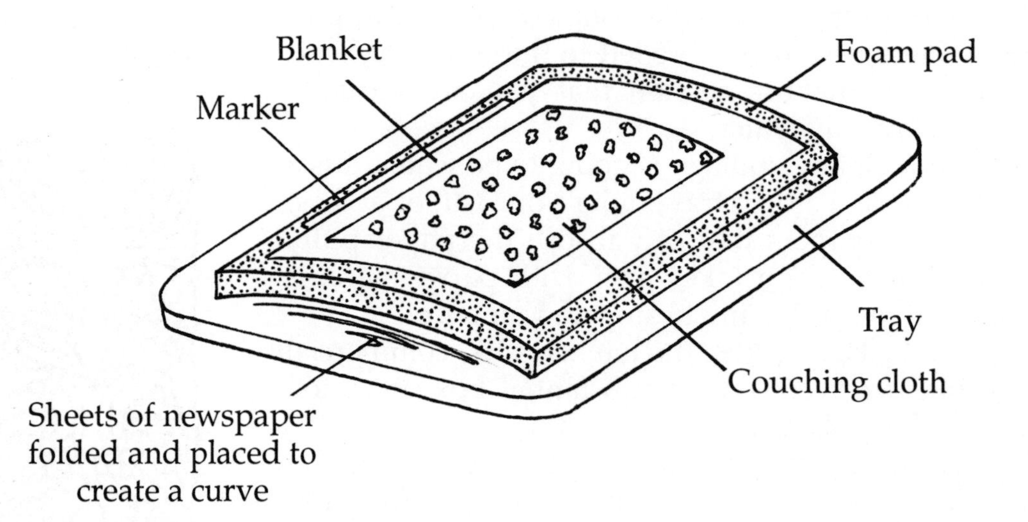

- Fold your towels or blanket to the appropriate size.
- Fold a newspaper lengthways into a flat roll about 5 to 7 cms (2" to 3") wide and place in down the centre of your tray. Fold four or five successive newspapers each to be a little wider than the previous one and layer them over the first newspaper, saturating each

with water before covering it with the next. You will have what looks like an upside down pyramid, with a curved top, down your tray centre.

- Saturate the foam and place this over the newspapers.
- Cover the whole thing with a saturated blanket piece.
- Lay down your marking strip on the edge nearest you, just where the bottom edge of the mould will rest before couching the pulp. You may like to practise first and see just where this strip should lay. When you are actually couching paper, this marker allows you to line up successive sheets of paper, all evenly on top of each other.
- Just before laying down your first piece of paper, cover the wet blanket with your first wet couching cloth, smoothed out to be wrinkle free, allowing the edge and corner of the marker to appear from under the nearside edge.

METHOD 2

To understand this method fully, take a look at the illustrations and instructions for *How to make a paper press,* showing you two ways of how to make your own simple press. Both presses do the same thing, and both, you will notice, involve upper and lower pressing boards. This method of couching uses these pressing boards.

- If you are making a press with wing nuts, place the two lower fastening planks down first, then place your lower pressing board onto these. Remember to work on a waterproof surface near your vat. You may have a basin large enough to place the planks and boards into. This is ideal, as all water run-off can be contained and your working area will remain fairly dry. You may have a baking dish large enough, or use a large kitchen tray.
- If you are making a press with G clips, simply place the lower pressing board onto a waterproof surface near your vat.
- Whichever press you choose, cover the board with a piece of saturated blanket. Lay down your marking strip, just where the bottom edge of the mould will rest before couching the pulp. You may like to practise first and see just where this strip should lay.

Remember, when you are actually couching paper, this marker allows you to line up successive sheets of paper, all evenly on top of each other.

- Lay a saturated couching cloth smoothed out to be wrinkle free onto the blanket, allowing the marker edge and corner to appear from under the nearside edge.

This lovely old Edwardian design could inspire you to make a sun-ray water mark

How to make a paper press

There are several ways of pressing your newly formed paper. They all involve boards, and can be as simple as a substantial breadboard with a member of your family standing on it! I'm quite sure you'll want to progress from this very simple method, though I assure you it works well, having made my first paper this way some years ago by convincing my youngest, large son to be the weight needed.

If you are fortunate enough to have an old bookbinder's press, you should use this. They are hard to find now, and would be expensive if you did locate one. They are very heavy as they're made from cast iron, and have a threaded screw that tightens down onto its base, pressing the contents very efficiently. The boards used in new presses need to withstand water, so choose marine ply or exterior ply of some sort. Other boards can be used, but they must be painted and sealed or they will swell and warp. If using new timber boards, saturate the surfaces with bleach and leave for about half an hour. The bleach will prevent the timber staining the paper. Neutralise the timber surface after washing off the bleach with a 50/50 solution of white vinegar and water, then rinse with fresh water. Dry the boards well and leave in an upright position to drain.

The presses illustrated both work on the same principle, and it's up to you whether you choose wing nuts or

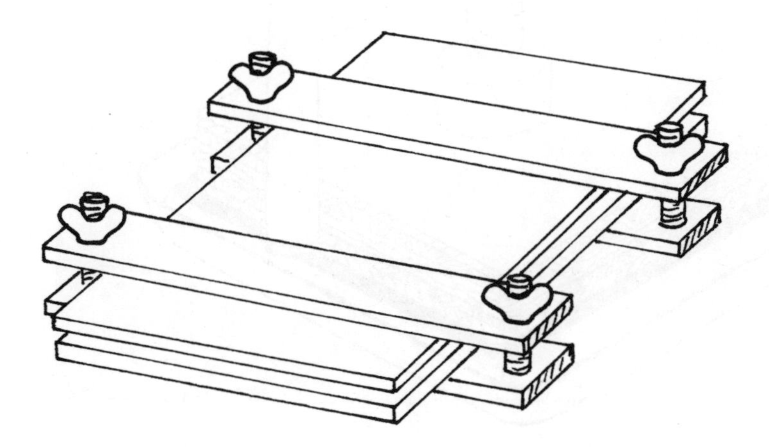

Press using wing nuts

G clamps as your fastenings. Make the press larger all round than your mould and deckle, and be sure, if using wing nuts, that the cross-timbers are strong enough not to bend too much during tightening, and therefore reduce the pressure in the centre of the press.

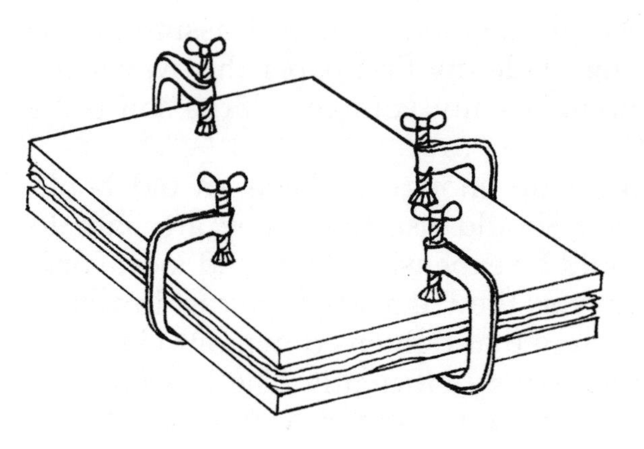

Press using G-clamps

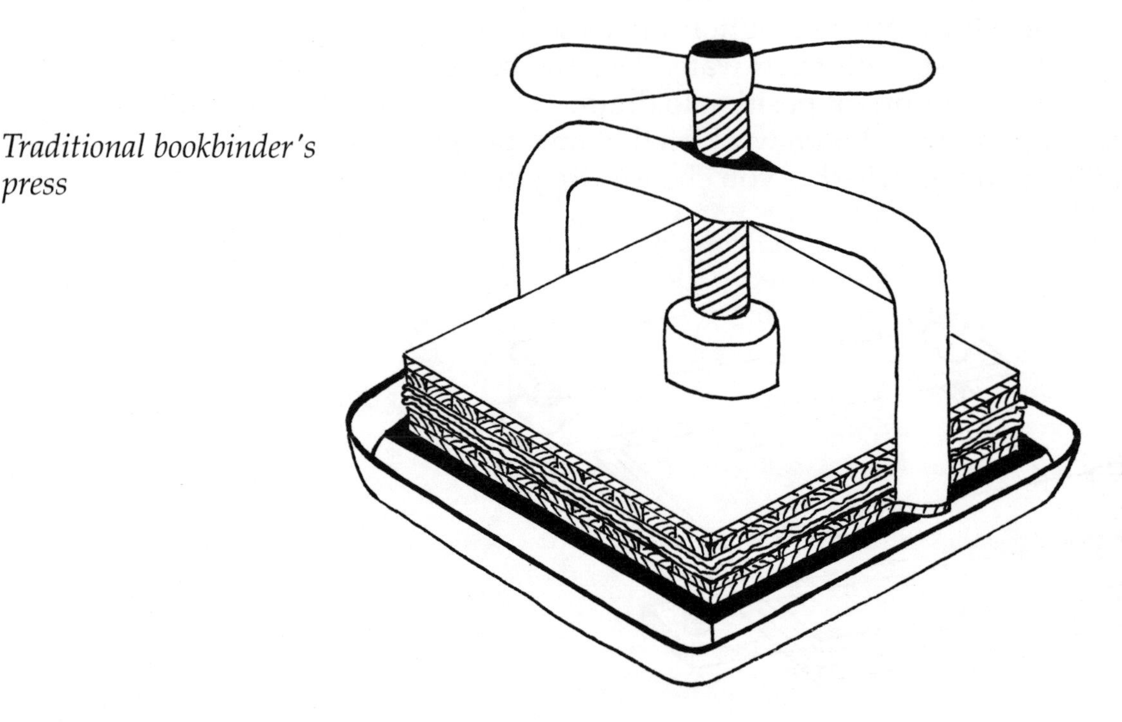

Traditional bookbinder's press

Making paper pulp

What papers are good for recycling?

Bear in mind that the better the paper is that you recycle, the better your end product will be. You will be faced with tremendous choices, as there are endless varieties of papers available as scrap. You must learn to be selective, as poor quality papers, such as newspapers, are not only difficult to reduce to clean pulp, but will produce poor quality paper.

Choose papers that were meant to be enduring in their original form. Wrapping papers and carry bags, envelopes, and computer, photostat, and writing papers are usually made to be strong and pliable. Their pulp will also be high quality and produce good papers. Some offices will supply you with bags of shredded documents that are excellent for pulping.

Avoid glossy magazine papers, sticky labels, waxy milk cartons or those from around frozen food, grey egg cartons or any paper that has a plastic, synthetic feel. Most commercial packaging that has been labelled 'recycled' is made from newspapers, and will not produce good paper. Try blending some recycled commercial packaging with other pulp for better results. You will not find it difficult to avoid these unsuitable paper types, as there is usually so much good paper available for recycling. Also, try to avoid combining papers of very different thicknesses, as the breakdown times are so variable.

Sort your papers into colours, if possible. It's useful to separate them into large transparent plastic bags, so that you can see at a glance what your supplies are like. Try, for example, to keep all blues and greens together, or all reds, yellows and oranges. Remember that if you combine two colours together your paper will result in the blending of these colours, such as blue and yellow pulp producing green paper. I like to sort through and keep any white scrap aside too, as it is needed to produce white or near-white paper, and is useful for softening the tone of brightly coloured pulp.

You should take care to discard any sticky surfaces, for they will not break down in the pulping process. Also remove from your scrap any staples, plasticised ridges, glitter from greeting cards and plastic windows from envelopes before tearing them up for pulping.

Tearing paper

Tear all your paper into pieces about the size of a postage stamp, or soak them in warm water then break them down by tearing the pieces apart. Keep a carton of scrap paper handy where the family gathers and you'll soon have plenty of willing hands tearing up paper. Paper that has been through the office shredder is ready to go straight into water after you've chopped it roughly with scissors, or you can break it down further under water.

Soaking the pieces

The length of soaking time will depend on your paper thickness. The finer the paper, the less time you'll need to soak it. Soaking paper pieces in warm water with a little dishwashing detergent added will hasten the saturating process. Buckets or similar containers are best for soaking the paper pieces, or use a garbage bin for large quantities.

You will only know if the paper has soaked long enough by the ease with which you are able to blend it. If large pieces still remain after blending, soak it longer. I have found that soaking paper overnight in warm water is usually enough for average-thickness scrap paper.

Pulping using the blender

As a general rule, half a cup of soaked paper should be mixed with 3 cups of water then added to the blender. Scoop it out of the soaking container with a seive and tip it into the blender after measuring it. You will soon know what half a cup of paper looks like!

A common problem with this method is the user's expectations of the blender. Too little water will not allow the blender to work well, and you'll have lots of large pieces of paper remaining in the pulp. Use your blender on high speed, taking care not to overheat it, and do observe the manufacturer's instructions about load levels. Be very careful of water near electricity. Keep plugs and switches dry, and mop up spills immediately. You should be able to hold a tumbler, full of diluted mixed pulp and water, up to the light and find that it is cloudy with minute particles of paper. Any large pieces remaining mean that it was not blended for long enough.

If you are using pulp immediately, pour it into the vat, and proceed with the next lot, half filling the vat. You can easily add more water to make up this depth. Remember that it is the pulp particles floating in the water that you will be collecting with your mould and deckle, and as water distributes the particles through the vat you can afford to be generous with added water. You can always add more strained pulp when the mixture thins.

If you are making pulp in advance, strain the pulp through your seive and discard the water. Place it in a sealed container in the refrigerator until it's needed. I like to add a drop or two of disinfectant if I'm storing it for any length of time. When you wish to make paper, add the strained pulp to the vat and pour in water until the vat is half full.

Wash your blender thoroughly before using it next for food preparation. The paper making process is non-toxic, and any residual particles will wash away easily.

Using an electric drill and paint stirring attachment

This attachment is a good compromise if no kitchen blender is available. Half fill a bucket with water and pulp then blend the mixture with the paint stirrer. Take care that the stirring attachment does not descend too far into the mixture so as to splash the drill with water, as wet electrical appliances are highly dangerous. As a safety measure, do not immerse the shaft of the stirrer any more than half to three quarters of its length into the water while blending. Blend for about 5 minutes then check the mixture using a tumbler, as for the electric blender method. Continue blending until all particles are minute. Transfer the mixture and add more water, if necessary, to half fill your vat.

Hand pulping

If no mechanical devices are available to you for making pulp, there is still the old stand-by of muscle power! A potato masher can, eventually, turn out quite respectable pulp when pounded constantly into a bucket or bowl of saturated paper pieces. Rubbing paper scrap against the kitchen grater will do much the same thing, but do take care not to damage your fingers on the grater. This is certainly working the hard way, but it is possible to produce acceptable pulp this way.

Making pulp from plant fibre

Most commercial paper today is made from wood chips, which are high in cellulose content. It is the cellulose content in all plant material that forms paper during the making process. Vast commercial plantations, and some natural forests of pine, eucalyptus and poplar trees throughout the world supply the needs of factories converting these wood chips into pulp for paper.

The home paper maker cannot efficiently pulp these tough wood chips, so other materials must be used. Leaves and stems of some fibrous plants are ideal for paper making, and often can be found growing in your garden. For example, banana leaves, palm fronds, lawn grass, gladioli and iris leaves, reeds and watsonia leaves are all suitable as they are fibrous and high in cellulose. Until you are a proficient paper maker, use only one plant fibre in any one batch of paper, or blend the plant fibre with some paper pulp. Combinations of plant fibre need experimentation, and are usually successful, but they are still a gamble in beginner's hands. I should caution you, before you go much further with plant based paper, about the breaking-down process. Leaves and stems are boiled in a mild caustic solution, and therefore, safety precautions must be observed. This process is not something children should undertake unsupervised. When using caustic soda, please follow these safety tips:

- Always wear protective eye glasses, and wearing rubber gloves should be considered.
- Keep a bottle of vinegar nearby for splashes on skin – the vinegar will neutralise the caustic soda.
- Add the caustic soda to the water, not the water to the caustic.
- In addition, follow the manufacturer's instructions on the container.

The process of breaking down the plant material into its basic fibres is caused by the action of the caustic soda. The soft spongy parts of leaves and stems are usually disposed of, leaving the stringy fibres that will make paper.

Cut up the plant material into small pieces about 2 cm (1") square, where possible discarding very tough parts. I suggest that you cut flax a little longer so that the long flax fibres can become a feature of the paper. Try to split the stems lengthways first, then cut across the leaf or stem to sever the fibres. Place the pieces into a stainless steel or enamel cooking pot and cover with cold water. Do not use aluminium pots or utensils. Add 1 teaspoon of caustic soda to each litre of water used. Some experts suggest that more caustic is necessary, and for tough plants this may be the case, but all agree that you should use as little caustic as possible.

Bring the pot to the boil slowly, stirring frequently with a wooden spoon. You will know when the plants are 'done' for the soft fibre becomes slimy and slides away from the structural fibres. Turn the heat off and prepare to rinse the fibres.

Tip the fibres into a sieve and dispose of the caustic water, adding some vinegar to it to neutralise the solution. Rinse the fibres under a running cold water tap, or tip them into a bucket and fill with water to rinse. Strain the fibres again and again until the water runs clear, then drain fibres thoroughly.

Add fibres to fresh water and blend into pulp. A mixture of half a cup of pulp to three cups of water seems adequate, but plant fibre will blend faster than paper pulp, so only blend for a matter of seconds before checking the process. The beauty of plant based paper is that the fibres show in the finished product, so don't blend the pulp to such an extent that these fibres are ground into nothing. If mixing it with paper pulp, or using it separately, add the pulp to water and half fill your vat.

You may like to keep some of the soft fibres aside before rinsing the pulp. I have found that these soft fibres can influence the colour of the finished paper. Simply rinse the fibres as described earlier – only allow some of the soft fibres to remain – then blend these and continue with the paper pulping process.

Soft plant fibres can make for very interesting colours when the plant fibre is combined with paper pulp to make paper. I suggest you experiment a little here.

Making sheets of paper

Your workplace

Try to set up your workplace to be as convenient as possible. You need to have immediate access to equipment for the various steps, and the thought of dripping water over the room while you walk from place to place is not a good one! So have all your bits and pieces on one bench, within reach – you'll probably find your paper making is not only easier, but more successful!

Remember to have your vat at a suitable height; consider the depth and your ability to reach into it. Your vat should be half filled with pulp and water mixture and it should be positioned with the long sides facing away from you. Have your container of extra pulp handy, and your couching pad on one side of the vat. Have plenty of towels to mop up splashes, and an extra jug of water for topping up the vat. Wet your mould and deckle and keep it nearby. I prefer not to wear rubber gloves while working – I find they get in the way, but you may find it easier.

Pulling sheets of paper

Agitate the pulp in the vat, until the particles are well distributed. Do this often during the pulling process, as the particles settle to the bottom very quickly.

Take the mould, mesh side uppermost, and place your deckle on top of it. Hold the mould and deckle together firmly, your hands on the short sides, with thumbs on top and fingers underneath.

This next move will come more easily with practice. Hold the mould and deckle almost vertically over the vat, and in one slow but steady motion, scoop the mould down into the water, slow down to be almost stationary in the water, then bring it out the other side holding the mould and deckle horizontally. What has happened during this action is this: when the mould and deckle were almost at a standstill, paper particles accumulated on top of the mesh, and were contained by the deckle. As you pulled the mould and deckle from the water, the force of the water draining from the mould pushed the particles down onto the mesh to bond together.

Immediately your mould and deckle leave the water, there will still be a shallow pool of water contained by the deckle. As this pool of water drains from the mould, very gently shake the mould and deckle from side to side, and from end to end to distribute the pulp more evenly. Do not do this after the water has drained or ridges may form in the pulp.

Lay the mould and deckle across one corner of the vat for a moment or two, until all dripping has ceased. Gently remove the deckle. It has done its job of containing the pulp to the desired paper size, and has given the paper the characteristic 'deckle edge' that is the traditional feature of handmade paper. As you remove it, take care not to allow any drips to fall from the deckle back onto the pulp; these will form craters and your paper will be spoiled. If this happens, simply place the mould back in the vat, wash off the pulp and pull another sheet.

Paper thickness is something you will become very aware of as your paper making progresses. When you pull your mould and deckle from the vat you should note the depth of the layer of particles. If you can see the mesh through the particles, chances are that your paper is too thin. As a beginner, thick paper is probably inevitable and will always be useable. Expert paper makers strive for thin papers, sometimes with a translucent quality. The ability to do this only comes with practice! If your paper is too thin, simply add more pulp to the vat; add more water if the paper is too thick.

If, for whatever reason, you are not happy with the paper sheet you have pulled, turn over the mould into the vat, tap the back and allow the pulp to wash off.

Steps in pulling a piece of paper

Step 1

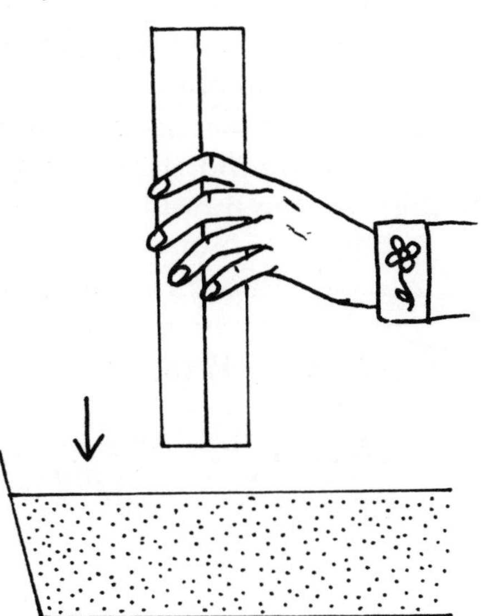

Step 2

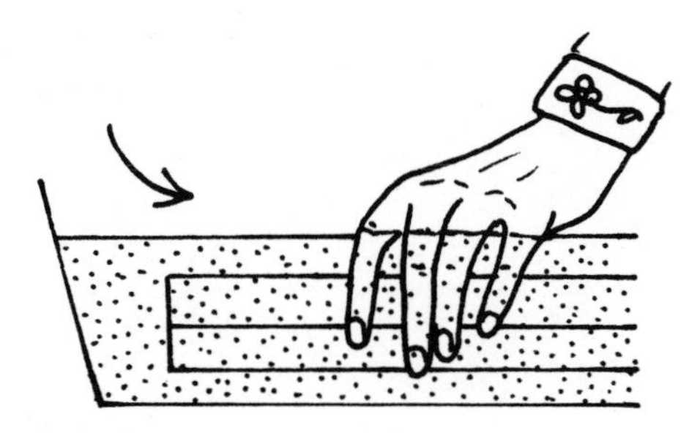

Step 3

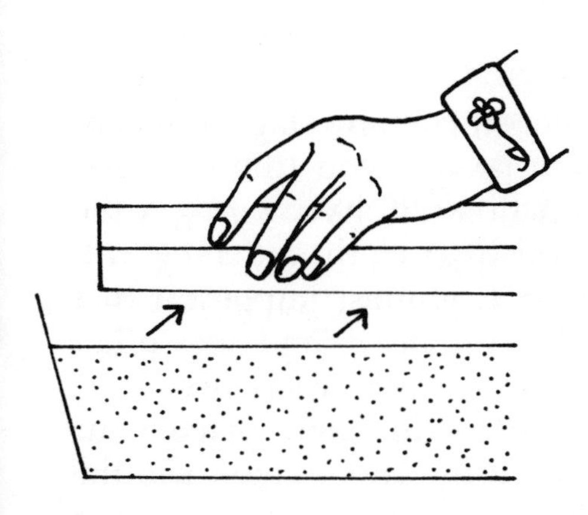

Step 4

Couching the paper

Take the mould, pulp side up, to the side of your couching pad. You will have placed the pressing board down covered with a wet blanket and couching cloth, if you choose this method, or you will have your newspaper and sponge rubber couch moist and with wet couching cloth on top, ready to receive the first paper. Check that your couching cloths are smooth and wrinkle free.

Hold the mould over the top of the couch and rest the nearside long edge on the marker accurately, lining up the corner of the mould with the end of the marker strip. Tip the mould to an angle away from you and check that the couching cloth is still in place and it has not adhered to the mould. Press the edge of the mould nearest you down into the couch firmly and roll the mould onto the

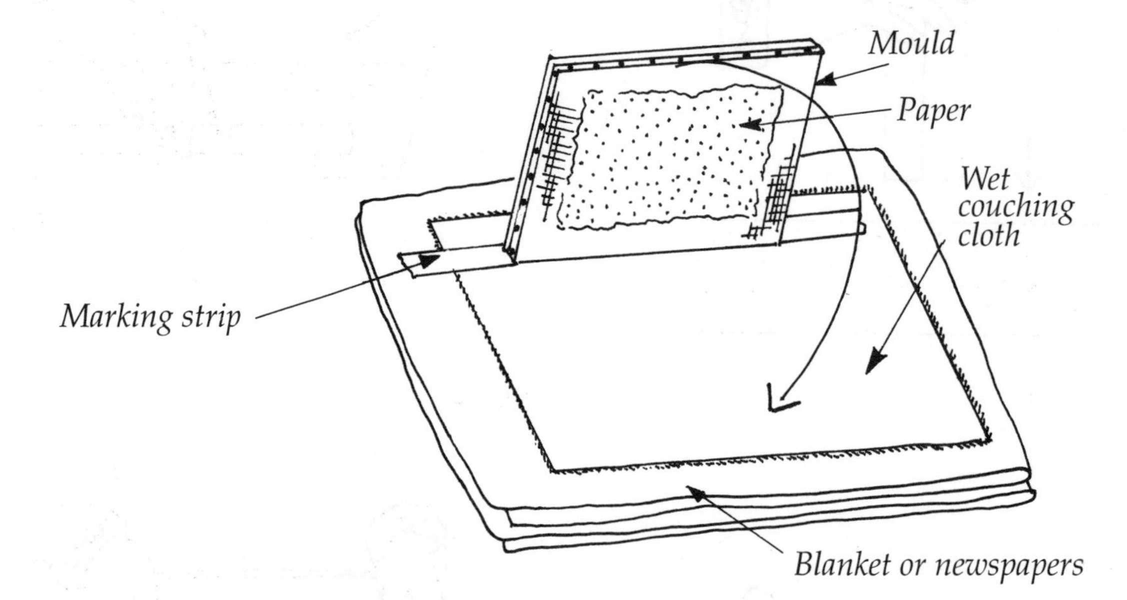

couch, pressing firmly. Do not stop at any stage, but continue by lifting off the edge nearest you (the one first placed on the couch). This action will have brought the pulp in contact with the couching cloth and they will have adhered to one another. It is most important that this motion is smooth and continuous, and like other craft skills, it will be bettered by further experience.

If you prefer, reverse the motion and place your marker on the opposite side of the couch, then roll the mould towards you pressing the pulp onto the couching

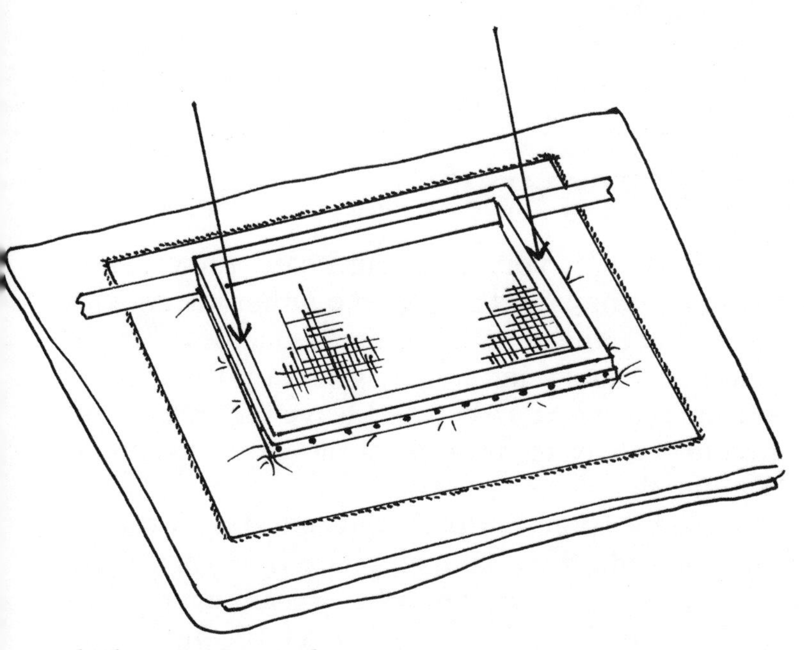

Push the mould firmly down onto the couch

cloth as instructed.

The first few sheets, inexplicably, often fail to couch well. Experts, too, have this problem, but they just proceed and pull more sheets. It is almost like seasoning your mould and deckle – and it soon rights itself and the couching becomes accurate. Cover the first sheet with another wet, smooth couching cloth. Repeat the pulling process and line up each successive paper sheet by using the marker. Build up a 'post' of papers, usually 6 to 10 sheets, then cover the top sheet with two couching cloths.

Don't couch papers of different dimensions onto the same post. The edges of the smaller ones will indent into the larger ones during pressing and spoil them. This also happens if paper is very badly aligned during couching.

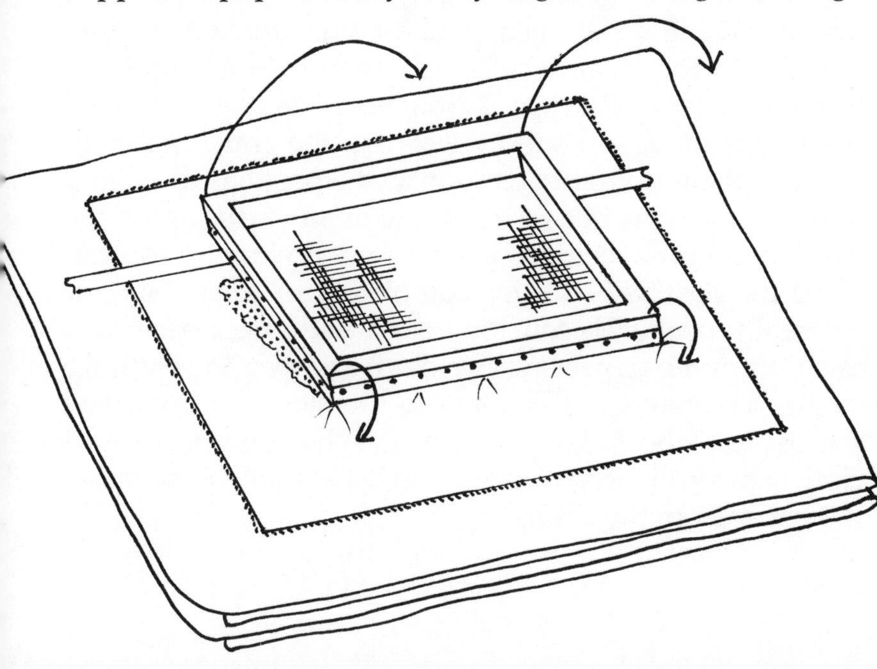

Lift the side of the mould that's closest to you off, pressing down on the front of the mould as you do so. This has brought the paper in contact with the couching cloth, allowing it to adhere

Pressing the paper

If you couched with your pressing board underneath, place the remaining board on top and either fasten with G clips, or place the remaining batons across the top board and install the wing nuts. Tighten up the clips or nuts until pressure is uniform, and leave the press at an angle to allow water to drain from the paper for about 10 to 15 minutes.

If you used a sponge rubber and newspaper couch, grasp the ends of all the couching cloths in each hand and transfer your post to the centre of the lower pressing board. Place the upper board over the post, and connect the fastening clips or nuts, as instructed earlier.

If after 10 to 15 minutes of pressing you find your post is still very damp, add a layer of newpapers top and bottom and re-press the paper. Pressing actually connects the particles or fibres together in a stronger bond than if the paper was not pressed. This is one aspect of paper making that I find fascinating. Paper, such as newspaper, that is made mechanically will have all its fibres pushed in the one direction (creating a 'grain') and subsequently tears easily along this direction. Handmade paper, due to the process of its creation, has fibres and particles settling in all directions, meshing together. Handmade paper will not tear at all easily.

Remove the upper pressing board and examine the paper to see if it finally looks dry, or at least only damp. I like to leave the post uncovered for a day or two to allow it to dry further. Damp paper is very vulnerable and any bump or nudge will dent it. Some experts like to remove the sheets of paper, still attached to the couching cloth and peg them up to dry as soon as the post has drained. I know this works, but an extra day or so of drying on the cloth in the post seems to prevent any accidents. You can leave the post as is, to dry completely, or after a day or so, gently peel each couching cloth with its paper sheet, away from the paper layer below it and peg the cloth up to dry. The paper will adhere to the cloth, so don't feel you have to be ready to catch it! When the paper and cloths are totally dry, unpeg them, as it's time to separate the paper from the cloths.

Separating paper from couching cloth

It will help you to get started if you grasp the cloth either side of the paper and pull gently. This eases the bond between paper and cloth at the top section, allowing you to gently prise back the top of the paper. With unhurried movements, peel the cloth from the paper, not the paper from the cloth. When all the paper is removed, stack it together, keeping the edges as even as possible, and press it again, using two dry couching cloths to protect top and bottom sheets from the pressing boards. If you have made some very textured sheets of paper, this last step may result in the adjoining sheets taking on the marks of the textured pieces. Interleave each sheet with dry couching cloths before pressing to avoid this problem.

Some paper makers then iron their paper, giving it a smooth, shiny surface, and have been known to use spray starch on very fine paper.

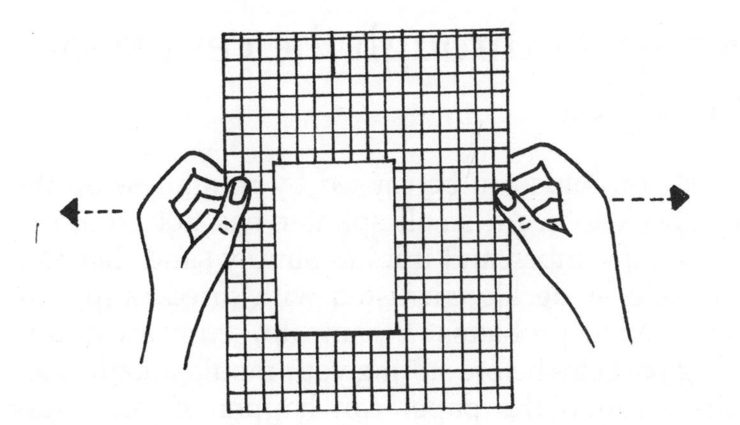

Separating the paper from the couching cloth

Common problems

Paper is too thin

This happens when there is insufficient pulp in the vat. Sometimes, in the excitement of making paper, you can forget to replenish the pulp. You can also be too quick in the pulling motion and fail to allow enough particles to settle onto the mould. Simply add more pulp to the vat to overcome this problem.

Paper is too thick

There is too much paper pulp in the vat. Take some out and dilute the remainder with water.

Paper splits through the centre, or at its edges

Edge splits or chips can be caused by roughness on the deckle. Check to see a small splinter has not crept out from the main timber, and if it has simply sand that area of the deckle. A deckle encrusted with dried pulp may also cause edge problems. Be sure to scrub away any remaining particles before storing your mould and deckle.

Splits through the paper can happen if you press very damp paper too quickly. The paper simply parts at the point of most pressure. Air trapped between layers can promote this too, so try a gentle pat with the hand on the top couching cloth, before the upper pressing board is fitted, to push out air bubbles.

Weak paper

Recycling poor quality paper often leads to this problem, and there is nothing much you can do about it. It can be given a stronger surface by dipping it in a gelatine solution (see sizing under the section headed *Creative effects*), or a good spray with starch prior to ironing it may help.

Paper can be weakened by drying it too quickly. Be sure it is only damp, not wet, if it is hung to dry, and that you press it enough to expel water. Remember, too, that pressing aids in the strength of paper, compressing the particles strongly together.

Pulp lumps

It is unlikely that you will find lumps of pulp at any stage after the vatting process. Mixing the pulp in water and pulling paper will surely reveal any impurities. However, if the gremlins strike and you are confronted with large pieces of pulp or fibre in your paper, you will have to put up with it, or tip the pulp back into the vat, then re-blend it all. Do not try to break down the lump while it's on your mould – you will destroy that piece of paper, and probably damage your mould.

Paper will not separate from couching cloths

This happens when paper is insufficiently dry before you try to separate the paper and cloth. The fine, fuzzy deckle edges can make it difficult to grasp the paper, but you should persevere. If the paper below adheres to the couching cloth above, leave it there to dry and separate it later, even if there is paper on the other side of the cloth.

Stains in paper

Stains can appear in your paper and are usually the result of something you have added which is leaching colour and staining the sheets above and below. Confetti is often a culprit. This is really only a nuisance and does not necessarily spoil the paper. Be sure to wash couching cloths well, as the stain will spread to more papers next time you use them. Bleach or strong washing detergent will remove the stains. I have found that colour from strongly coloured pulp can stain couching cloths and therefore stain any further papers. Strong washing action usually removes colour and any remaining coloured fibres from the cloths.

Spots of solid colour, and transparent spots

Solid coloured paper that has been recycled sometimes does not blend completely, and appears as small dots of colour over the paper. Some people strive for this effect, others dislike it. Blend your pulp more to avoid this, and, to be sure, smear a little undiluted pulp over glass and allow it to dry. You will soon see the effect it will give.

Flower petals included in your plant fibre pulp can, for some reason, dry transparent. This can appear as a hole in your paper, when in fact, it is a transparent layer. You should be aware that these spots left by flower petals will eventually perish to leave a small hole. This may, or may not, create a problem.

Dry paper has creases, or indentations

Sometimes your couching cloth is pushed into creases during the couching process. This can happen if you push your mould across it, picking up the paper, instead of pressing the mould onto the cloth. Try to move the mould only in a downwards rolling motion that pushes the cloth and paper down, not across. Creases can also be caused by threads from the edge of your couching cloth being pressed between layers of paper. Avoid this problem by finishing off the edges of your cloths to prevent fraying.

Straightening wrinkled dry paper

Dampen a couching cloth and spread it over a pressing board. Interleave paper with damp cloths until they are all in a post, smoothing them with your hand as you lay on the next couching cloth. Install the top pressing board and tighten the press. Check after an hour or so, as the wrinkles may disappear in this short time. Press for longer if necessary. Spread the paper out to dry completely, then iron it.

Creative effects

You will find that any decorative bits added to the pulp will naturally float to the top of your paper pulp on the mould. This leaves one side of your paper – that closest to the mesh – fairly plain, which is good for the writing surface. Bear in mind that any bulky additions may make writing difficult, but will look wonderful anyway.

Additions to the pulp

Grasses and seeds

Add these as the pulp is poured into the vat, or at the blending stage if they are bulky. Only blend them for a very short time, or they will be unrecognisable.

Tissue paper

Torn up lightweight tissue paper added to the blender will create very interesting paper. Wet strength paper napkins do not break down easily, and so remain in recognisable particles. Be sure they are torn into very tiny pieces prior to blending. Do tear up these tissue papers, as cut edges look peculiar.

Threads

Wind sewing machine thread (or even fine knitting yarn) around your finger many times and snip through the loops, cutting it into many tiny pieces. Add this to the pulp in the vat and you will have a fascinating surface effect on your paper. These threads tangle with the pulp and remain recognisable. For a really special effect snip up lengths of metallic sewing machine thread, or trim the sparkles off Christmas tinsel to include in paper for Christmas cards.

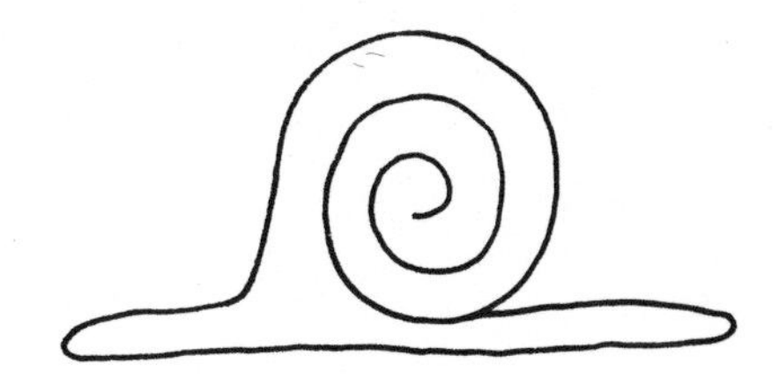

Confetti

You can make your own using a hole punch, and that way you can choose your colours. Add the pieces to the vat while pulling paper.

Flowers

Gather fresh flowers and tiny, curled leaves such as parsley and dispose of any bulky, hard parts; petals and leaf tips are best. Blend them in the blender with clear water for only a few seconds. Add this to the vat and then pull your paper. You can be very generous with added flowers, or only add a few. Remember that flower petals can dry transparent in the paper.

Watermarks

These have the effect of leaving an indentation in the pulp, creating a motif or initials, that is noticeable because it is thinner and the light shines through it and emphasises it. It is a great way of personalising your paper, or making a very special gift. Bend fuse wire into the desired shape; a small simple motif, or your initials. Securely stitch this shaped wire onto one corner of your mould using strong sewing thead, or use lengths of fine wire to secure the shape. Remember to reverse initials – consider the side of the paper that they will appear on and position them accordingly. Pull your paper as usual, and couch it. When it's dry you will notice the watermark when you hold up your paper sheet to the light.

Embossing

This makes an indentation in the wet paper. Form a shape from cardboard, or use a flat leaf or coin and place down onto the couching cloth in readiness for the pulp from the next paper pull. The object will leave its indentation in the paper, and will pop out when the paper is dry and removed from the couching cloth.

Some simple shapes suitable for watermarks

Colouring the paper

Some vegetable dyes do not have sufficient strength to fully colour a vat of paper, though they will have a subtle effect. The dyes in coloured paper napkins are excellent and quite vibrant, so pulp these by themselves and add this pulp to the vat. You will have to blend these for some time, as the napkins are made to resist breaking up in water. Powdered paint, such as that used by kindergarten children, is a good colourant, and it is usually non-staining. Some fabric dyes are terrific, but they will stain your couching cloths, your mould, your deckle – in fact, everything you use! You'll get good results, but do consider the cost.

Duplex sheets

This is an exciting process. Pull a thin sheet of paper, and couch it. Lay a leaf, flowers or whatever you choose on top of the sheet. Do make sure the object is flat! Pull another sheet and couch it directly over the first sheet, enclosing the object. Press the sheet in the normal way and you'll end up with a pretty, thickish paper that has an enclosed object! Consider enclosing cut-outs from fabric or small decorative pieces cut from paper. If you want to go a step further, make your second sheet of paper from another batch of coloured pulp. The trick is to have both sheets line up at the edges, especially if you are using pulps of two colours, and this will only come with practice.

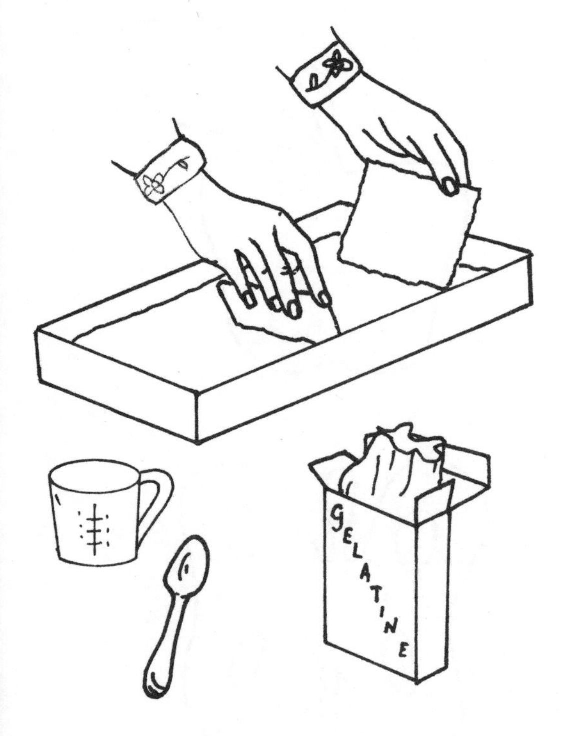

Sizing with gelatine

Sizing

This seals the surface of your paper and makes it suitable to receive water based paints or inks. In short, it waterproofs your paper. Without sizing, these paints will run into the paper fibres, and if you are planning to use your paper for calligraphy, this step is necessary. Allow new paper to cure for about 2 weeks before sizing it. Dissolve $2^1/_2$ teaspoons of gelatine in 1 litre (32 fl oz) of warm water, and pour this solution into a shallow container, such as a baking dish, that is larger than your paper. As paper tends to break up when wet, you may have to experiment here. Holding a sheet at both ends, slide it through the gelatine solution and out, angling it to drain off excess solution. Place the sheet flat onto glass or plastic sheets and allow to dry.

Or, during the paper making process, add a handful of starch to cold water, then tip it into the vat. This fills the paper somewhat, and also gives it a suitable surface for writing. Trial and error are necessary to arrive at suitable quantities of added starch per vat of pulp and water.

Paper patchwork

This is an interesting, permanently visible way of using your lovely handmade papers. While just about any patchwork block pattern would be adaptable, you should select shapes that will be easy to cut, and show off the deckle edges of the paper. I explain the patchwork photographed in the captions. Aim to combine colours that blend, and consider using the patterned papers along with plain ones. I have used papers that were made from various leaves and other vegetation, giving interesting depths of colour and tones that blend very well with each other.

Simply cut out trial shapes of commercially made paper and experiment first, seeing what pieces should overlap, or butt up to each other, and whether you need a background piece of paper. You can cut out pieces from your handmade paper later once you are happy with the composition. Using a background piece of paper can be of great benefit, giving you a base, and an extra colour, at the same time. The background piece should be as large as the complete pattern, and it is advisable to use an acid-free paper, usually made from rags, as it will be as enduring as your handmade paper. Remember that it is the edges of the paper which give this craft its particular charm.

Apply glue to the back of the handmade paper using a small paintbrush. Do not use excess glue, as this will cause bumps and possible wrinkles. I like to sometimes use a mixture of 75 per cent Clag (school glue) and 25 per cent Aquadhere (PVA glue), otherwise I use straight PVA glue sparingly. These glues are non-toxic and should not stain the paper. Have your paper patchwork framed when completed, using a clear glass as a cover.

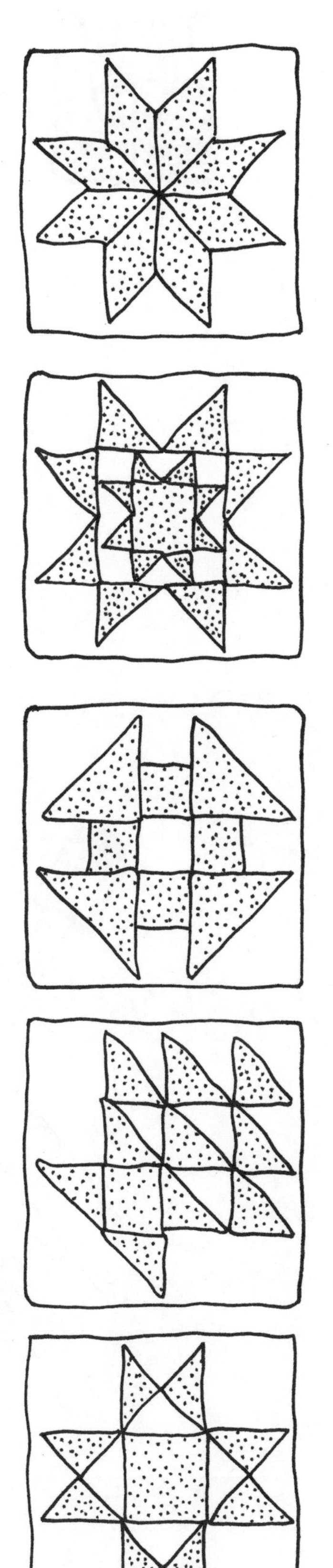

Paper patchwork

Simple marbling

This craft is a good one for adults and children alike. My instructions are for the simplest method though far more complex methods exist. There are many examples of fine marbling carried out by craftspeople, especially in antique books, where you will find endpapers marbled. Europe, especially Italy, is the home of marbling, where the craft is still strongly practised.

The process of marbling is simple, but the more you do, the better you become. The patterns formed in the first instance will soon show you where you are going wrong – your results are immediately obvious. A wide toothed comb drawn in zigzags across the surface of the bath will give you an antique look, and very thinned paint will give a different result from thicker paint. When you have finished with a paint combination, you can lift off the paint from the glue surface with strips of paper towel, leaving the glue ready to be used again, and to receive another layer of paints.

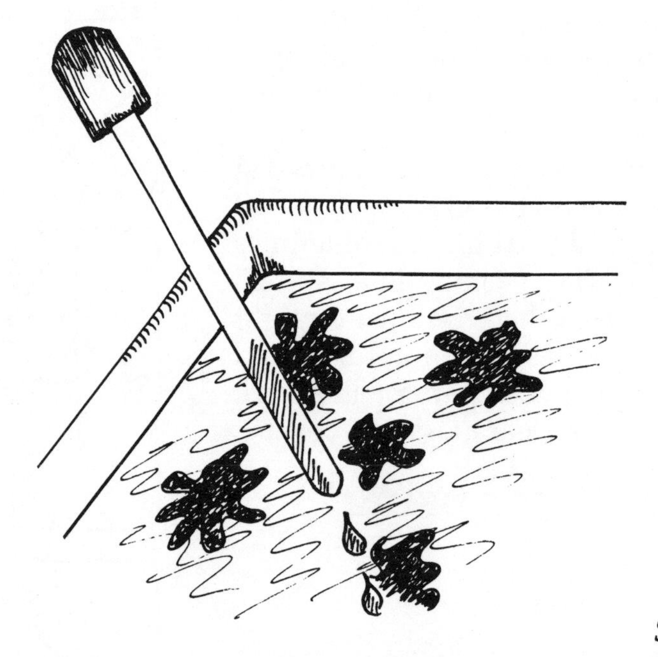

Step 1. Adding drops of oil paint to the surface of the wall-paper glue

MATERIALS

- Handmade paper to be decorated
- Tubed oil paints
- Paint thinners such as white spirit or turpentine
- Paper towel
- Wallpaper glue
- Ox gall (available from artist's supply shops)
- Large baking dish
- Eye dropper

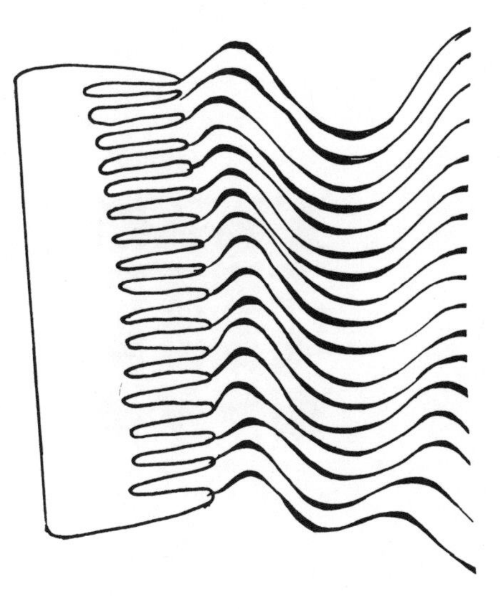

Step 2. Drag the comb over the surface of the glue, distributing the paint in wavy lines

METHOD

- Mix wallpaper glue according to manufacturer's instructions. Half fill the baking dish with glue.
- Mix paints to a liquid consistency, adding a few drops of Ox gall to each colour. Using an eye dropper, distribute these colours as desired across the surface of the glue bath forming patterns as you go.
- Take a sheet of paper, which fits within the baking dish, and, holding it at either end, place one end of paper onto the surface of the glue, dropping the other end gradually down onto the surface. Lift paper from the starting point, working quickly and not allowing paper to be re-dipped. You can re-dip the paper to build up another layer of paint, but this must be uniform, not an accident!
- Lay paper flat to dry, then press it between weights if you need to flatten completely.

Creative suggestions for using marbled paper

- Use your marbled paper to cover notebooks, tins to hold pencils and pens, and cover pencils – the ends will sharpen as usual and the paper will be cut away by the blade of the pencil sharpener. You will have

to use your thinner sheets of paper for these areas that need to be flexible. Ordinary PVA glue is ideal for fastening handmade paper, but take care that you don't leave residue on the paper surface, as it dries glossy.

- Marble your handmade paper envelopes, and dip the end of paper sheets, or one corner, into the marbling paints to make matching writing paper.
- Remember that marbled paper needs to be totally dry before it can be used.
- Cover a wooden or papier maché box with sheets of marbled paper. The larger areas of box, such as the lid and sides, could be covered with butted pieces of paper, forming a patchwork-like effect. Glue the paper directly onto the box, being sure it's flat, and trimming overlapping paper away from the edges with a sharp razor-blade or scissors.

Stencilling your handmade paper

Stencilling onto your beautiful handmade paper and envelopes is a good way of decorating them. It is simple to make your own stencils, and, even if you have not sized your paper, it is still possible to 'paint' it using oil stencil crayons.

I have illustrated a number of possible stencil motifs, and they're sized to fit into corners of paper, and onto the closing flaps of envelopes. Select which ever design you prefer, and proceed to cut your stencil.

Gather around you the following equipment: a sheet of manilla cardboard, linseed oil and turpentine, a small jar, some soft scrap fabric, a sharp craft knife or scalpel, soft lead pencils, a cutting surface (see page 41), your choice of paints – either artist's acrylics or oil stencil crayons, stencil brushes or small natural sponges and your handmade paper to be decorated.

First, mix up a small quantity of 50 per cent turpentine and 50 per cent linseed oil in a screw-top jar and shake the mixture well. Take a fabric scrap and saturate it with the mixture. Wipe it well into a piece of manilla

cardboard that is larger than the stencil design you've selected. Allow it to dry for 15 minutes or so, then wipe away any excess oiliness. Dampen and dispose of the oily cloths for safety.

Place the stencil design against a window, and using this make-shift 'lightbox' place your oiled cardboard over it and trace off the design using a lead pencil, cross-hatching those areas to be cut out. If you have a patchworker's self-healing cutting board, use it to cut out the sections of the design from the cardboard, using a scalpel, craft knife or similar. Breadboards make good bases for cutting upon, as do wads of evenly placed newspapers.

Position the stencil over the area of the paper you wish to decorate. If you are using acrylic paints on sized paper, pour only very little acrylic paint into a flat container, and dab your stencil brush or natural sponge into it. Dab off excess paint onto a scrap of kitchen paper towel. Start to fill in the design with paint, using light dabbing motions with your brush or sponge. Don't try to make the effect opaque, as a little unevenness in paint application is charming. Continue stencilling more motifs as desired.

If you are using oil stencil crayons, rub the end of a crayon onto scrap paper, leaving an area of colour large enough to rub your brush into. While the brush is being rubbed into the colour, particles of 'paint' are being

attached to the bristles. Note that no liquid of any sort is used in the application of oil crayons. Lightly apply the brush over the stencil, using gentle circular motions, building up colour until you're satisfied with the paint coverage. Oil stencil crayons give a light, misty coverage of the paper and look quite different to painted stencils. Brushes can be cleaned with mineral turpentine after use with oil crayons.

I have another use for stencils, using oil crayons on handmade paper, that may appeal to you. Stencil a more complicated design onto a larger sheet of paper and then frame your work. The handmade paper makes an intriguing background to the stencil, and the finished article is sure to be a conversation starter. A series of these framed stencils along a wall can look wonderful.

Glossary for paper making

Papyrus

The reed-like plant used by the ancient Egyptians to make paper. Papyrus grows close to water along river banks, and is still used today for making beautiful, enduring papers.

Mould

The lower section of the combined 'mould and deckle'. Mesh is fastened across the mould and it is on this that the paper pulp rests during the paper making process

Deckle

The upper section of the combined 'mould and deckle'. The depth of the deckle helps determine the thickness of the finished paper, and its size dictates the dimensions of the paper. The mould and deckle are of identical sizes, with the deckle, if possible, having less depth than the mould. Traditional writing paper often has a soft wavy edge called a 'deckle edge'.

Couch

As the name implies, this is the place where the freshly pulled paper is put to rest! Couches can be very simple flat pads made from old towels or blankets, or slightly more complex with a curved surface to provide drainage for the still-wet paper.

Pulling paper

The term given to the general process of making a sheet of paper. If you have inserted your mould and deckle into your vat filled with paper pulp and water and pulled it out, you are said to have 'pulled a sheet of paper'.

Vat

The trough or container used to hold the water and paper pulp. The vat should be larger all round than your mould and deckle and roomy enough to allow you to sweep the mould and deckle through it unhindered. Your kitchen sink can also become your vat.

Couching cloths

These are interleaved between the freshly couched sheets of paper. I prefer to use cotton fabric such as calico or sheeting, but commercially made kitchen wipes can be substituted. Be sure if using fabric couching cloths to oversew the edges to prevent stray threads marring the surface of your paper.

Marker strip

A narrow strip of fabric placed on the top of your couch to help you align successive sheets of paper.

Post

This is the term used to describe a pile of freshly pulled paper.

Watermark

A motif shaped from wire that is applied to the mesh on the mould. The image of this motif left on the dried paper is known as a watermark.

Paper Press

Various types of presses are suitable, and, if your are a beginner, a press can be as simple as two heavy boards. The paper press works to expel unwanted moisture from the freshly pulled paper, and the pressing action helps to amalgamate the fibres of the paper into a strong bond.

Sizing

This occurs when dried paper is passed through a gelatine solution, rendering it suitable for caligraphy and paint application.

Projects shown on colour pages

Page 1

Stencilled writing paper

I've used stencil crayons for the charming motif on the corners of the writing paper. The crayons are soft and powdery, and when applied, leave a mist of colour rather than an opaque layer of paint. The paper's undulating surface did not seem a problem when stencilling.

Page 2

Assorted papers

Paper can be charming when it is presented as a scroll – perhaps you could used an old-fashioned wax seal to close it! Envelopes are folded to take advantage of the lovely uneven paper edges, and scrap paper and plant papers are stacked in an old Georgian tea caddy.

Page 3

Ingredients for paper making

These are just some of the plant fibres you can use to make paper, and most can be found in domestic gardens. Flax, mulberry, golden elm, convolvulus (known as morning glory), onion skins and banana leaves are all readily available. Shredded paper was happily given to me from the office, and tissue papers and brown papers are easily salvaged from everyday use.

Page 4

Machine-embroidered papers

These are not exceptional examples of precise machine

embroidery, rather they are the freehand rambles of a curious sewer – me! First, I chose sheets of paper that were thicker than usual. The finer papers will tear easily under the machine needle. I lowered the feed dogs on my Bernina, installed the darning foot (a foot with a small round hoop and a spring action) and threaded the bobbin with gold thread. My upper thread was ordinary sewing thread. When the feed dogs are lowered you have control over the movement of the paper, and the machine foot does not actually hold the paper down. Guide the paper through the machine slowly so that the needle stitches the pattern you wish to make. A minute dab of glue to hold trimmed threads will secure your stitching.

Page 5

Monkey wrench paper patchwork

The edges of handmade paper have always appealed to me, and I felt some patchwork may be a new way to take advantage of this lovely effect. Heavy deep pink cardboard was soaked for days before pulping. Large squares of the pink were made and the best selected (an envelope sized mould and deckle was used). I divided the square into six, to ascertain the size of the triangles I would need to cut. I then cut the centre square and the number of triangles required from banana leaf paper. Using minute dabs of PVA glue – or stick glue (it won't soak through to the right side) – I first fastened the square to the centre. The triangles were then applied, working outwards from the centre. You will note that the spacing right at the edges is not even – but it doesn't seem to matter. I would think any patchwork block that required straight-sided pieces would be suitable for paper patchwork, so experiment to find other interesting designs.

Page 6

Plant paper patchwork

I don't believe this is a traditional patchwork design – it seemed to evolve because the paper edges suggested the shapes! All these papers are made from plants. I actually glued these to a sheet of rag-based paper (acid free

paper is available from art supply shops, and will be very enduring as a backing sheet). The sheet of backing paper was slightly smaller all round than the completed patchwork piece. Working from the centre outwards, the papers are: mulberry in the centre square, flax adjoining the square, with banana leaf adjoining the flax. The large green triangles are convolvulus (morning glory), with small mulberry triangles resting on large triangles of brown onion skin paper.

Page 7

Using a mould and deckle

This is my favourite old mould and deckle. I can't imagine how many pieces of paper this pair have made! It's a little unconventional as I covered the deckle with a very strong cheesecloth type fabric, which obviously has some synthetic content as it has lasted so long. This fabric surface gives my paper a rag-like look, which I feel is rather nice. The paper has just been drained, I've lifted off the deckle and will proceed to couching the paper.

Page 8

A watermark

You can have some fun experimenting with these. My feeling is that they need to be fairly simple to be effective, and thicker, rather than thinner, to make any impression. Remember to place it on the opposite side of the deckle to where you wish it to appear on the paper. It would not be appropriate to fasten a watermark to a fabric-covered deckle, as the wire would probably tear the fabric.

Couch and couching cloths

This is the simplest couch of all. Experts may not like the simple nature of the flat blanket acting as a couch, as it does not really provide drainage. However, it works well for beginners, and I suggest you try it. To make matters even simpler, the pine breadboard that it is resting on makes it easy to transport the stack of couched papers outside, where you can place another heavy board on top, then weight it – or even stand on it – to press the water from the stack. This method of couching and pressing is so simple it's worth a try, before you actually make a press for yourself. The couching cloths have all had their edges neatened, to prevent stray threads being embedded in the paper. In this photograph, I have not used a marker to line up the stack of papers being couched. There are times when the brightness of the paper colour shows through and acts as its own marker, and it's easy to align the next sheet exactly!

A pot of flax

Chopped up domestic flax being cooked in water, with a dash of caustic soda. Plants are extremely varied in the amount of cooking time they take to break down. Flax and convolvulus take the longest, in my experience. Most others, such as banana leaves, quickly become pulpy and mushy. The easiest way to check on progress is to remove a leaf and rinse it under cold running water. If the green matter slides away easily it is probably ready. Some experts suggest discarding all the pulp, and just using the fibres. I disagree, as the pulp can become a colouring agent. Convolvulus is a good example. I had just about given up trying to achieve green paper that dried as green as it was when wet, as all plant fibre papers tended to turn pale earthy colours when dried. I added some of the plant pulp to the mixture of brown scrap paper and plant fibres ($^1/_3$ paper, $^2/_3$ plant fibre) and green paper appeared!

Credits

Thank you to Grosvenor Antiques, Lindfield, Sydney for the Georgian tea caddy, the Regency apprentice's cupboard and the pine apprentice's box on the cover; to DMC/Myart in Sydney for their stencil crayons; and to Bernina for my wonderful sewing machine.

For information about craft kits and materials available from Tonia Todman, please write to:

Tonia Todman Craft Kits
PO Box 12
Balmain NSW 2041
Australia

Notes

NOTES

NOTES